FRESHWATER AQUARIUM PLANNER & LOGBOOK

A Little Planner to Help You Maintain Your Freshwater Aquarium

WALTER JAMES

Cover Design by Jared Sandoval.

The Icons used in this work were designed by:

- Freepik, Smashicons, Becris, Dimi Kazak, monkik, macrovector, djvstock, pikisuperstar, pch.vector, studiogstock, rawpixel.com

Published by Admore Publishing: Berlin, Germany

Printed in the United States of America

www.admorepublishing.com

Contents

Foreword

There is something truly magical about the fish-keeping hobby.

... But besides looking nice, does caring for fish actually add to your life?

It absolutely does! Fish tanks have so many benefits for those who own and interact with them. Studies have shown that owning fish has positive physiological and psychological effects. Having a fish tank improves your mood and reduces stress and anxiety. Observing fish elegantly float around their homes for just a few minutes lowers your heart rate and blood pressure.

There are many benefits to owning fish and maintaining an aquarium beyond this, but with all these advantages, however, comes some responsibility. Fish help us in so many ways (half of which we don't realize immediately), so it is essential that we return the favor with our care for them.

This book will help you with some of the basics of maintaining your freshwater aquarium. It will also help you plan out it's unique look and allow you to keep track of when you completed important tasks like a water test or a water change. If you are looking for more in depth knowledge in regards to freshwater aquariums consider my other book "Freshwater Aquariums for Beginners". That covers everything you will need to properly care for a freshwater aquarium, and it's scaly inhabitants. This way, your flappy little guy or girl will live a very long, happy, and healthy life.

Now, let's get started with your planner!

Life is better with a fishtank!

ONE

Let's get Down to Basics

Little compares to admiring your own aquarium: the tranquility, the fun and exciting inhabitants, the beauty of nature. It's incredible that we can all own a little slice of wonder like this in our homes.

Aquariums do require some maintenance work, however.

When did I last do a water change? Did I already do a water test this week, or was that last month?

This planner will help you keep track of your tasks and gives a little advice on the side. Many of us (*ahem myself especially*) dive pretty deep into the hobby and end up owning several tanks. Then the whole keeping track of everything process can get even more overwhelming. This book hopes to simplify the process and plan things out properly so that you and the inhabitants of your aquarium are happy.

Tanks

Size

Since I'm new to fish keeping and aquariums, I'll just start with a small one. This way, I can get used to things, and they are less work to maintain.

The above statements may seem like a good idea, and many beginners think this, but it is far better to start out with a larger tank. Small tanks get dirty extremely fast, which means you will spend more time cleaning and doing water changes than you would for a larger tank. Additionally, freshwater aquarium fish will need more room than you think.

My recommendation is for those new to the hobby to start with a tank size of around 20 gallons. This size will allow you to experiment with various looks and fish, and you won't need to make great financial investments.

Regarding what fish you choose to keep, it is recommended to do individual research depending on the species. Different species will need different amounts of space. In general, however, it is recommended to have at least one gallon of water for every inch of fish. So, If you have a very beginner-friendly 20-gallon tank, you could stock it with up to 20 fish that reach one inch in length when fully grown. Alternatively, you could select 10 fish of 2 inches, 6 fish of 3 inches, and so on.

Remember, fish can be much smaller than they will grow up to be when they are first purchased. Consider how big they will be when they are fully grown when making a purchasing decision. If you are ever unsure, opt for more space than you think you will need.

Aquatic animals are generally happier with more room to swim freely and explore.

Location

There are many suitable places to keep an aquarium in your home, however, there is a primary factor to keep in mind. Don't keep the tank in any areas that will interfere with the temperature of the tank. Having the temperature of the tank fluctuate constantly will harm the inhabitants. Consider all of the following:

- Areas where direct sunlight hits the tank to an extent where temperature can rise too high.
- Keeping a tank near windows may also be unsuitable. If you live in a cold climate or extremely warm, having a tank too close to a window will affect water temperature.
- Keeping a tank near heaters, air vents, or similar apparatus will also be a risk for temperature fluctuations.

It is also critical to keep your tank on a stable and robust surface. Tanks filled with water are HEAVY. Make sure that what it is sitting on can carry serious weight.

Equipment

This section will cover the essential equipment you will need to keep your aquarium inhabitants happy and healthy. They perform various tasks that help keep the environment healthy and allow you to create a beautiful underwater world.

Filter

A filter will help keep the water in your tank clean and creates a healthy biome for your aquarium animals. There are a variety of filter styles ideally suited to any aquarium. The most popular and beginner-friendly filters are called **H**ang **O**n **B**ack filters (HOB filters).

Heater

Many of the fish and water animals you will look to add to your aquarium will come from tropical regions. In order to create a suitable home for them in your home, you will typically need a heater to match the temperature. There are many types of heaters. Many can now measure a specific temperature and so keep a stable climate, preventing any temperature spikes.

Thermometer

It is excellent to keep a thermometer. It will help you perform water changes and allow you to quickly double-check the temperature of your aquarium's water.

Light

You will want to make sure the aquarium has a good light source so you can adequately see your lovely aquarium. Furthermore, if you have live plants in your tank, it will allow them to flourish and grow. LED lights are generally a great choice.

Plants

Plants do not only look good and are an excellent way to decorate your tank, but they can also provide more comfort to your finned roommates. Some fish enjoy being able to rub or hide in leaves. Having plants allows them to feel right at home. You can opt for live plants, or you can opt for artificial plants. If you decide on artificial plants, check on quality, as sometimes they can have sharp edges that end up harming some fish's fins.

Substrate

Substrate is the material that can be found at the bottom of your tank. There is an incredible variety of substrates available. Gravel, sand, rock, just to name a few. Here it is essential to research what type of environment the fish or inhabitants you would like to get come from and need.

Décor

These are the items you use to decorate your tank. Just like plants, décor can help make your aquarium look beautiful and create a safe and happy home for your fish. Many fish enjoy having lots of hiding places to explore, such as plants, caves, pots, driftwood, or rocks. Let your creative juices flow.

Siphon

A vacuum style siphon helps you clean your tank and helps you perform water changes. It is a great tool that allows you to get into the substrate and hard to reach places of your tank. A definite must-have!

Bucket

Buckets are always helpful to have close by. You never know when there is an emergency, and you need to remove water from your tank or your fish. A bucket is excellent for when you need to perform water changes as well.

Water Tests & Water Conditioner

Pick up some water tests to check on your tank's water quality. They help you check on everything from nitrates and nitrites to ammonia and pH values. Also consider purchasing some water conditioner. This allows you to convert tap water into properly treated aquarium water. Depending on where you live, tap water may contain harmful chemicals such as chlorine. A water conditioner can clear out these chemicals and allows you to provide safe water for your aquarium.

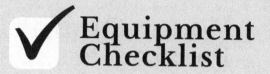# Equipment
Checklist

- ☐ Filter
- ☐ Heater
- ☐ Thermometer
- ☐ Light
- ☐ Plants
- ☐ Substrate
- ☐ Décor
- ☐ Siphon
- ☐ Bucket
- ☐ Water Tests & Water Conditioner

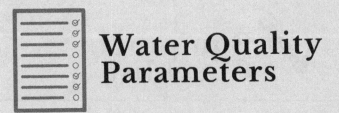

Water Quality Parameters

Temperature	75°–80° F
pH	7.0-7.8
KG	4°–8°
GH	4°–10°
Ammonia	0.0
Nitrite	0.0
Nitrate	<50 ppm
Phosphate	<0.5 ppm

*Note: These are generalized figures. Your tank and its inhabitants may require different water conditions.

You are
o<u>ffish</u>ially
the best!

TWO

Maintenance

Now that we have considered the basics and have our equipment in order, what are my tasks to maintain my tank?

There are a variety of things you can do to ensure that everything is A-Ok. Below I have created little maintenance tips that are split up into different time frames. Some tasks are performed daily, some weekly or bi-weekly, some monthly, and some periodically.

This may sound overwhelming at first, but trust me, it is very simple and a lot of fun. It also simplifies the process of maintaining your aquarium. Enjoy!

Maintenance Schedule

Daily Tasks

Observe fish & plants (health)

Check if equipment is fuctioning

Temperature check

Feed fish (remove un-eaten food)

Write down potential concerns

Weekly/bi-weekly Tasks

Wipe down/vacuum/remove algae

Remove dead leaves

Partial water change (15-25%)

Perform a water test

Maintenance Schedule

Monthly Tasks

Perform a water test

Trim plants

Remove debris

Larger water change (25-50%)

Periodical Tasks

Replace old equipment (ex. light bulbs)

Fertilize plants

Exchange substrate

<u>Warning:</u>
I may start talking about aquariums at any time.

THREE

Performing A Water Change

Performing a water change is one of the primary ways of maintaining your aquarium. It removes everything from dangerous chemicals or bacteria to uneaten fish food and waste. Not performing a water change at least bi-weekly can significantly lower water quality and turn your beautiful aquarium into an unsafe place for its inhabitants.

Aim to perform a water test weekly and particularly look at the values for nitrite, nitrate, and ammonia. If these figures are nearing or in the danger zone, perform a 50% water change as soon as possible. Having these values out of order is highly dangerous for your fish and aquarium life. If all the figures on a water test are fine, you can skip a week. However, you should aim to perform a water change bi-weekly or at a minimum, once a month, depending on your tank.

Perform a water change by following the following steps:

- Unplug all electrical items:

Unplug your heater, filter, or any other electrical items you may have in your aquarium. These items can be damaged or break down if the water level of your tank lowers significantly. Many are not meant to run outside of water, and it can cause system breakdowns.

- Interior wipe down:

Before removing any water, with a clean cloth, wipe down the interior glass of your aquarium. It is common for a variety of particles to build up on the glass of your tank. Wiping down the sides, loosen up any dirt or build-up that you can then remove during the water change.

- Grab your siphon:

Use a siphon to perform your water change. This will allow you to suction up dirt and debris at a controlled pace and reach difficult places within the tank. Make sure to get deep into substrate where dirt may collect that you cannot see at first glance. Aim to remove anywhere between 25-50% of the aquarium's water during a water change. Be careful not to remove over 50% of the water, as this can actually be very counterproductive to keeping a healthy tank.

- Refill the tank:

Once you have removed the appropriate amount of water and suctioned up the dirt, you can start refilling the aquarium. Here, it is crucial that the freshwater you add to the tank is the same temperature as the water already in your aquarium. Getting this wrong can shock your fish and is very unhealthy.

Also, treat your water before adding it to the tank with a water conditioner. Use a water conditioner that is appropriate for your tank and follow the instructions on the label. Treating water from your tap will remove various chemicals that may be present and make it safe for your fish. Slowly pour the treated and appropriate temperature water into your tank. Do this carefully as you don't want to mess up the ecosystem and environment of your tank.

Plug back in all of the electronics, and you are done with your water change. As a final touch, it is also recommended to clean the outside glass of your tank. Make sure if you use chemicals that you do not accidentally get any of it inside of your tank. This can be very dangerous for your fish and other water animals. You can otherwise opt for an eco cleaner or simply use paper towels.

Filter Tips

Your filter will also need to be maintained and cared for to ensure a healthy ecosystem in your tank. It is great to give it some attention once a month, but make sure to never give it a thorough cleaning the same day you perform a water change. The filter contains many healthy bacteria and keeps your aquarium thriving. If you complete a water change and clean your filter simultaneously, it can remove too many of these beneficial bacteria and chemicals and leave your tank to contaminate. Try to wait around 5 days after a water change before cleaning your filter.

Cleaning a filter mainly consists of rinsing it in aquarium water. Depending on what filter you have, you may need to:

- Replace cartridges or pads. Filters with these items are not meant to last much longer than a month.
- Rise the sponge, bio media, and ceramic. Make sure not to rinse these items in fresh tap water but instead in the aquarium itself.
- Look through your filter's instructions and study how the user's manual states you should clean the apparatus.

*Ask me about
my fishtank!*

FOUR

Aquarium Planner

PLAN OUT THE LOOK OF YOUR AQUARIUM(S)

Use this section to plan out the look of your tank or tanks. Do some research for what you would like to include, this be the type of substrate, plants, decorations, and of course what kinds of inhabitants. Although the sample designs are rectangular, you can still plan out your aquarium regardless of the shape it may have.

Have fun and be creative. It can be as basic or as complex as you like, but it always works best when you have a plan...

TANK: _Bedroom   3-29-22_

Notes:

3 fish - goldfish -
1" white 2" gold & white
3" gold & white
Maggie, Jane, Glen

TANK: _____

Notes:

_____

_____

_____

_____

_____

_____

_____

_____

_____

TANK: _____

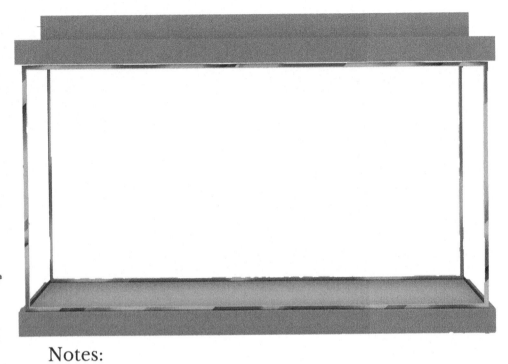

Notes:

_____
_____
_____
_____
_____
_____
_____
_____
_____

TANK: _____

Notes:

_____
_____
_____
_____
_____
_____
_____
_____
_____

TANK: _____

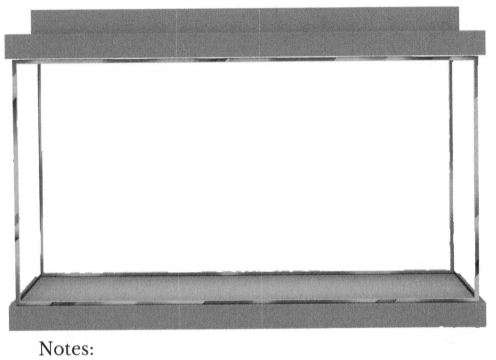

Notes:

_____

_____

_____

_____

_____

_____

_____

_____

_____

_____

TANK: _____

Notes:

_____

_____

_____

_____

_____

_____

_____

_____

TANK: _____

Notes:

_____

_____

_____

_____

_____

_____

_____

_____

_____

TANK: _____

Notes:

_____
_____
_____
_____
_____
_____
_____
_____
_____

TANK: _____

Notes:

_____
_____
_____
_____
_____
_____
_____
_____
_____

TANK: _____

Notes:

_____
_____
_____
_____
_____
_____
_____
_____
_____

Just one more fish,
I promise!

FIVE

Maintenance Log

Use this section to keep track of all your maintenance habits. This will make it so much easier to remember when you last performed a water change or a water test.

It will also help you keep track of any changes you make to various tanks and make notes on any differences you noticed that you want to keep your eye on.

Bedroom

Tank: _____ 3-27-22 – Sunday _____ Date: _____

GENERAL CHECK LIST:

Check Equipment [✓]   Remove Debris [✓]

Water Test ? [✓]   Trim Plants [✓]

Rinse Filter [✓]   Water Change | 0% | 15% | (25%) | 35% | 50% |

results didn't line up w/ gauge —

WATER PARAMETERS:

Temperature __ [?]

pH _____ [ ]

KG _____ [ ]

GH _____ [ ]

Ammonia ____ [ ]

Nitrite _____ [ ]

Nitrate _____ [ ]

Phosphate ____ [ ]

FISH & ANIMALS:

Feeding

once daily flakes

Behavior

eager

Additions

Lost Animals

NOTES

Water change simpler than I
thought. Let water get warm
bfore putting it in ~~the~~ tank.
looks unclear (?)
Siphon easy. – Do I want
another tank? – Beta = ?
catfish? Guppies?

Tank: Bedroom -
3-29-22 Tuesday **Date:**

GENERAL CHECK LIST:

Check Equipment ☑   Remove Debris ☑

Water Test ☑   Trim Plants ☒

Rinse Filter ☐   Water Change | 0% | ⊗ | 25% | 35% | 50% |

WATER PARAMETERS:

Temperature __ ☑

pH ____ OK __ 6.8

KG _____ 2.

GH ____ OK __ 150

Ammonia ____ 2.

Nitrite _ high ☑ 200

Nitrate _ high ☑ 10

Phosphate OK 6.2

FISH & ANIMALS:

Feeding

Behavior

Additions

Lost Animals

NOTES

High nitrite
High Nitrate

Tank: *2½ gallon - Bedroom*   **Date:**

GENERAL CHECK LIST:

Check Equipment ☑   Remove Debris ☐

Water Test ☐   Trim Plants ☐

Rinse Filter ☐   Water Change | 0% | 15% | 25% | 35% | 50% |

WATER PARAMETERS:

Temperature__ ☐

pH_____ ☐

KG_____ ☐

GH_____ ☐

Ammonia____ ☐

Nitrite_____ ☐

Nitrate_____ ☐

Phosphate____ ☐

FISH & ANIMALS:

Feeding

Behavior

Additions

Lost Animals

NOTES

Tank: **Date:**

GENERAL CHECK LIST:

Check Equipment ☐   Remove Debris ☐

Water Test ☐   Trim Plants ☐

Rinse Filter ☐   Water Change | 0% | 15% | 25% | 35% | 50% |

WATER PARAMETERS:

Temperature __ ☐

pH _____ ☐

KG _____ ☐

GH _____ ☐

Ammonia ____ ☐

Nitrite _____ ☐

Nitrate _____ ☐

Phosphate ____ ☐

FISH & ANIMALS:

Feeding

Behavior

Additions

Lost Animals

NOTES

Tank: _____ **Date:** _____

GENERAL CHECK LIST:

Check Equipment ☐   Remove Debris ☐

Water Test ☐   Trim Plants ☐

Rinse Filter ☐   Water Change | 0% | 15% | 25% | 35% | 50% |

WATER PARAMETERS:

Temperature __ ☐

pH _____ ☐

KG _____ ☐

GH _____ ☐

Ammonia ____ ☐

Nitrite _____ ☐

Nitrate _____ ☐

Phosphate ____ ☐

FISH & ANIMALS:

Feeding
☐

Behavior
☐

Additions
☐

Lost Animals
☐

NOTES

Tank: _____ **Date:** _____

GENERAL CHECK LIST:

Check Equipment ☐   Remove Debris ☐

Water Test ☐   Trim Plants ☐

Rinse Filter ☐   Water Change | 0% | 15% | 25% | 35% | 50% |

WATER PARAMETERS:

Temperature__ ☐

pH_____ ☐

KG_____ ☐

GH_____ ☐

Ammonia____ ☐

Nitrite_____ ☐

Nitrate_____ ☐

Phosphate____ ☐

FISH & ANIMALS:

Feeding

Behavior

Additions

Lost Animals

NOTES

Tank: _____   **Date:** _____

GENERAL CHECK LIST:

Check Equipment ☐   Remove Debris ☐

Water Test ☐   Trim Plants ☐

Rinse Filter ☐   Water Change | 0% | 15% | 25% | 35% | 50% |

WATER PARAMETERS:

Temperature__ ☐

pH_____ ☐

KG_____ ☐

GH_____ ☐

Ammonia____ ☐

Nitrite_____ ☐

Nitrate_____ ☐

Phosphate____ ☐

FISH & ANIMALS:

Feeding

Behavior

Additions

Lost Animals

NOTES

Tank: **Date:**

GENERAL CHECK LIST:

Check Equipment ☐   Remove Debris ☐

Water Test ☐   Trim Plants ☐

Rinse Filter ☐   Water Change | 0% | 15% | 25% | 35% | 50% |

WATER PARAMETERS:

Temperature__ ☐

pH_____ ☐

KG_____ ☐

GH_____ ☐

Ammonia____ ☐

Nitrite_____ ☐

Nitrate_____ ☐

Phosphate____ ☐

FISH & ANIMALS:

Feeding

Behavior

Additions

Lost Animals

NOTES

Tank: _____   **Date:** _____

GENERAL CHECK LIST:

Check Equipment ☐   Remove Debris ☐

Water Test ☐   Trim Plants ☐

Rinse Filter ☐   Water Change | 0% | 15% | 25% | 35% | 50% |

WATER PARAMETERS:

Temperature__ ☐

pH_____ ☐

KG_____ ☐

GH_____ ☐

Ammonia____ ☐

Nitrite_____ ☐

Nitrate_____ ☐

Phosphate____ ☐

FISH & ANIMALS:

Feeding

_____

Behavior

_____

Additions

_____

Lost Animals

_____

NOTES

Tank: **Date:**

GENERAL CHECK LIST:

Check Equipment ☐   Remove Debris ☐

Water Test ☐   Trim Plants ☐

Rinse Filter ☐   Water Change | 0% | 15% | 25% | 35% | 50% |

WATER PARAMETERS:

Temperature _ _ ☐

pH _ _ _ _ _ _ _ _ _ ☐

KG _ _ _ _ _ _ _ _ _ ☐

GH _ _ _ _ _ _ _ _ ☐

Ammonia _ _ _ _ ☐

Nitrite _ _ _ _ _ _ ☐

Nitrate _ _ _ _ _ _ ☐

Phosphate _ _ _ _ ☐

FISH & ANIMALS:

Feeding

Behavior

Additions

Lost Animals

NOTES

Tank: **Date:**

GENERAL CHECK LIST:

Check Equipment ☐   Remove Debris ☐

Water Test ☐   Trim Plants ☐

Rinse Filter ☐   Water Change | 0% | 15% | 25% | 35% | 50% |

WATER PARAMETERS:

Temperature __ ☐

pH _____ ☐

KG _____ ☐

GH _____ ☐

Ammonia ____ ☐

Nitrite _____ ☐

Nitrate _____ ☐

Phosphate ____ ☐

FISH & ANIMALS:

Feeding

Behavior

Additions

Lost Animals

NOTES

Tank: **Date:**

GENERAL CHECK LIST:

Check Equipment ☐   Remove Debris ☐

Water Test ☐   Trim Plants ☐

Rinse Filter ☐   Water Change | 0% | 15% | 25% | 35% | 50% |

WATER PARAMETERS:

Temperature__ ☐

pH_____ ☐

KG_____ ☐

GH_____ ☐

Ammonia____ ☐

Nitrite_____ ☐

Nitrate_____ ☐

Phosphate____ ☐

FISH & ANIMALS:

Feeding

Behavior

Additions

Lost Animals

NOTES

Tank: **Date:**

GENERAL CHECK LIST:

Check Equipment ☐   Remove Debris ☐

Water Test ☐   Trim Plants ☐

Rinse Filter ☐   Water Change | 0% | 15% | 25% | 35% | 50% |

WATER PARAMETERS:

Temperature__ ☐

pH_____ ☐

KG_____ ☐

GH_____ ☐

Ammonia____ ☐

Nitrite_____ ☐

Nitrate_____ ☐

Phosphate____ ☐

FISH & ANIMALS:

Feeding

Behavior

Additions

Lost Animals

NOTES

Tank: _____ **Date:** _____

Check Equipment ☐   Remove Debris ☐

Water Test ☐   Trim Plants ☐

Rinse Filter ☐   Water Change | 0% | 15% | 25% | 35% | 50% |

WATER PARAMETERS:

Temperature _ _ ☐

pH _ _ _ _ _ _ _ _ ☐

KG _ _ _ _ _ _ _ _ ☐

GH _ _ _ _ _ _ _ _ ☐

Ammonia _ _ _ _ ☐

Nitrite _ _ _ _ _ _ ☐

Nitrate _ _ _ _ _ _ ☐

Phosphate _ _ _ _ ☐

FISH & ANIMALS:

Feeding

Behavior

Additions

Lost Animals

NOTES

Tank: **Date:**

GENERAL CHECK LIST:

Check Equipment ☐   Remove Debris ☐

Water Test ☐   Trim Plants ☐

Rinse Filter ☐   Water Change | 0% | 15% | 25% | 35% | 50% |

WATER PARAMETERS:

Temperature__ ☐

pH_____ ☐

KG_____ ☐

GH_____ ☐

Ammonia____ ☐

Nitrite_____ ☐

Nitrate_____ ☐

Phosphate____ ☐

FISH & ANIMALS:

Feeding

Behavior

Additions

Lost Animals

NOTES

Tank: _____ **Date:** _____

GENERAL CHECK LIST:

Check Equipment ☐   Remove Debris ☐

Water Test ☐   Trim Plants ☐

Rinse Filter ☐   Water Change | 0% | 15% | 25% | 35% | 50% |

WATER PARAMETERS:

Temperature__ ☐

pH_____ ☐

KG_____ ☐

GH_____ ☐

Ammonia____ ☐

Nitrite_____ ☐

Nitrate_____ ☐

Phosphate____ ☐

FISH & ANIMALS:

Feeding
☐

Behavior
☐

Additions
☐

Lost Animals
☐

NOTES

Tank: _____ **Date:** _____

GENERAL CHECK LIST:

Check Equipment ☐   Remove Debris ☐

Water Test ☐   Trim Plants ☐

Rinse Filter ☐   Water Change [0%] [15%] [25%] [35%] [50%]

WATER PARAMETERS:

Temperature_ _ ☐

pH_ _ _ _ _ _ _ _ _ ☐

KG_ _ _ _ _ _ _ _ _ ☐

GH_ _ _ _ _ _ _ _ _ ☐

Ammonia_ _ _ _ ☐

Nitrite_ _ _ _ _ _ ☐

Nitrate_ _ _ _ _ _ ☐

Phosphate_ _ _ _ ☐

FISH & ANIMALS:

Feeding

Behavior

Additions

Lost Animals

NOTES

Tank: **Date:**

GENERAL CHECK LIST:

Check Equipment ☐   Remove Debris ☐

Water Test ☐   Trim Plants ☐

Rinse Filter ☐   Water Change | 0% | 15% | 25% | 35% | 50% |

WATER PARAMETERS:

Temperature __ ☐

pH _____ ☐

KG _____ ☐

GH _____ ☐

Ammonia ____ ☐

Nitrite _____ ☐

Nitrate _____ ☐

Phosphate ____ ☐

FISH & ANIMALS:

Feeding

Behavior

Additions

Lost Animals

NOTES

Tank: **Date:**

GENERAL CHECK LIST:

Check Equipment ☐   Remove Debris ☐

Water Test ☐   Trim Plants ☐

Rinse Filter ☐   Water Change | 0% | 15% | 25% | 35% | 50% |

WATER PARAMETERS:

Temperature __ ☐

pH_____ ☐

KG_____ ☐

GH_____ ☐

Ammonia____ ☐

Nitrite_____ ☐

Nitrate_____ ☐

Phosphate____ ☐

FISH & ANIMALS:

Feeding

Behavior

Additions

Lost Animals

NOTES

Tank: **Date:**

GENERAL CHECK LIST:

Check Equipment ☐   Remove Debris ☐

Water Test ☐   Trim Plants ☐

Rinse Filter ☐   Water Change | 0% | 15% | 25% | 35% | 50% |

WATER PARAMETERS:

Temperature__ ☐

pH_____ ☐

KG_____ ☐

GH_____ ☐

Ammonia____ ☐

Nitrite_____ ☐

Nitrate_____ ☐

Phosphate____ ☐

FISH & ANIMALS:

Feeding

Behavior

Additions

Lost Animals

NOTES

Tank: **Date:**

GENERAL CHECK LIST:

Check Equipment ☐   Remove Debris ☐

Water Test ☐   Trim Plants ☐

Rinse Filter ☐   Water Change | 0% | 15% | 25% | 35% | 50% |

WATER PARAMETERS:

Temperature __ ☐

pH _____ ☐

KG _____ ☐

GH _____ ☐

Ammonia ____ ☐

Nitrite _____ ☐

Nitrate _____ ☐

Phosphate ____ ☐

FISH & ANIMALS:

Feeding

Behavior

Additions

Lost Animals

NOTES

Tank: _____ **Date:** _____

GENERAL CHECK LIST:

Check Equipment ☐   Remove Debris ☐

Water Test ☐   Trim Plants ☐

Rinse Filter ☐   Water Change | 0% | 15% | 25% | 35% | 50% |

WATER PARAMETERS:

Temperature __ ☐

pH _____ ☐

KG _____ ☐

GH _____ ☐

Ammonia ____ ☐

Nitrite _____ ☐

Nitrate _____ ☐

Phosphate ____ ☐

FISH & ANIMALS:

Feeding
☐

Behavior
☐

Additions
☐

Lost Animals
☐

NOTES

Tank: _____ **Date:** _____

GENERAL CHECK LIST:

Check Equipment ☐ Remove Debris ☐

Water Test ☐ Trim Plants ☐

Rinse Filter ☐ Water Change | 0% | 15% | 25% | 35% | 50% |

WATER PARAMETERS:

Temperature__ ☐

pH_____ ☐

KG_____ ☐

GH_____ ☐

Ammonia____ ☐

Nitrite_____ ☐

Nitrate_____ ☐

Phosphate____ ☐

FISH & ANIMALS:

Feeding

Behavior

Additions

Lost Animals

NOTES

Tank: _____   **Date:** _____

GENERAL CHECK LIST:

Check Equipment ☐   Remove Debris ☐

Water Test ☐   Trim Plants ☐

Rinse Filter ☐   Water Change | 0% | 15% | 25% | 35% | 50% |

WATER PARAMETERS:

Temperature __ ☐

pH _____ ☐

KG _____ ☐

GH _____ ☐

Ammonia ____ ☐

Nitrite _____ ☐

Nitrate _____ ☐

Phosphate ____ ☐

FISH & ANIMALS:

Feeding

Behavior

Additions

Lost Animals

NOTES

Tank: _____   **Date:** _____

Check Equipment ☐   Remove Debris ☐

Water Test ☐   Trim Plants ☐

Rinse Filter ☐   Water Change [0%] [15%] [25%] [35%] [50%]

WATER PARAMETERS:

Temperature __ ☐

pH _____ ☐

KG _____ ☐

GH _____ ☐

Ammonia ____ ☐

Nitrite _____ ☐

Nitrate _____ ☐

Phosphate ____ ☐

FISH & ANIMALS:

Feeding

Behavior

Additions

Lost Animals

NOTES

Tank: _____   **Date:** _____

GENERAL CHECK LIST:

Check Equipment ☐   Remove Debris ☐

Water Test ☐   Trim Plants ☐

Rinse Filter ☐   Water Change [0%] [15%] [25%] [35%] [50%]

WATER PARAMETERS:

Temperature__ ☐

pH_____ ☐

KG_____ ☐

GH_____ ☐

Ammonia____ ☐

Nitrite_____ ☐

Nitrate_____ ☐

Phosphate____ ☐

FISH & ANIMALS:

Feeding

Behavior

Additions

Lost Animals

NOTES

Tank: _____   **Date:** _____

GENERAL CHECK LIST:

Check Equipment ☐   Remove Debris ☐

Water Test ☐   Trim Plants ☐

Rinse Filter ☐   Water Change | 0% | 15% | 25% | 35% | 50% |

WATER PARAMETERS:

Temperature__ ☐

pH_____ ☐

KG_____ ☐

GH_____ ☐

Ammonia____ ☐

Nitrite_____ ☐

Nitrate_____ ☐

Phosphate____ ☐

FISH & ANIMALS:

Feeding

Behavior

Additions

Lost Animals

NOTES

Tank: _____   **Date:** _____

GENERAL CHECK LIST:

Check Equipment ☐   Remove Debris ☐

Water Test ☐   Trim Plants ☐

Rinse Filter ☐   Water Change | 0% | 15% | 25% | 35% | 50% |

WATER PARAMETERS:

Temperature__ ☐

pH_____ ☐

KG_____ ☐

GH_____ ☐

Ammonia____ ☐

Nitrite_____ ☐

Nitrate_____ ☐

Phosphate____ ☐

FISH & ANIMALS:

Feeding
☐

Behavior
☐

Additions
☐

Lost Animals
☐

NOTES

☐

Tank: **Date:**

GENERAL CHECK LIST:

Check Equipment ☐   Remove Debris ☐

Water Test ☐   Trim Plants ☐

Rinse Filter ☐   Water Change | 0% | 15% | 25% | 35% | 50% |

WATER PARAMETERS:

Temperature _ _ ☐

pH _ _ _ _ _ _ _ _ ☐

KG _ _ _ _ _ _ _ _ ☐

GH _ _ _ _ _ _ _ _ ☐

Ammonia _ _ _ _ ☐

Nitrite _ _ _ _ _ _ ☐

Nitrate _ _ _ _ _ _ ☐

Phosphate _ _ _ _ ☐

FISH & ANIMALS:

Feeding

Behavior

Additions

Lost Animals

NOTES

Tank: _____ **Date:** _____

GENERAL CHECK LIST:

Check Equipment ☐   Remove Debris ☐

Water Test ☐   Trim Plants ☐

Rinse Filter ☐   Water Change | 0% | 15% | 25% | 35% | 50% |

WATER PARAMETERS:

Temperature__ ☐

pH_____ ☐

KG_____ ☐

GH_____ ☐

Ammonia____ ☐

Nitrite_____ ☐

Nitrate_____ ☐

Phosphate____ ☐

FISH & ANIMALS:

Feeding

☐

Behavior

☐

Additions

☐

Lost Animals

☐

NOTES

Tank: _____   **Date:** _____

GENERAL CHECK LIST:

Check Equipment ☐   Remove Debris ☐

Water Test ☐   Trim Plants ☐

Rinse Filter ☐   Water Change | 0% | 15% | 25% | 35% | 50% |

WATER PARAMETERS:

Temperature __ ☐

pH _____ ☐

KG _____ ☐

GH _____ ☐

Ammonia ____ ☐

Nitrite _____ ☐

Nitrate _____ ☐

Phosphate ____ ☐

FISH & ANIMALS:

Feeding

Behavior

Additions

Lost Animals

NOTES

Tank: _____ **Date:** _____

GENERAL CHECK LIST:

Check Equipment ☐   Remove Debris ☐

Water Test ☐   Trim Plants ☐

Rinse Filter ☐   Water Change | 0% | 15% | 25% | 35% | 50% |

WATER PARAMETERS:

Temperature _ _ ☐

pH _ _ _ _ _ _ _ _ ☐

KG _ _ _ _ _ _ _ _ ☐

GH _ _ _ _ _ _ _ _ ☐

Ammonia _ _ _ _ ☐

Nitrite _ _ _ _ _ _ ☐

Nitrate _ _ _ _ _ _ ☐

Phosphate _ _ _ _ ☐

FISH & ANIMALS:

Feeding

Behavior

Additions

Lost Animals

NOTES

Tank: _____   **Date:** _____

GENERAL CHECK LIST:

Check Equipment ☐   Remove Debris ☐

Water Test ☐   Trim Plants ☐

Rinse Filter ☐   Water Change | 0% | 15% | 25% | 35% | 50% |

WATER PARAMETERS:

Temperature__ ☐

pH_____ ☐

KG_____ ☐

GH_____ ☐

Ammonia____ ☐

Nitrite_____ ☐

Nitrate_____ ☐

Phosphate____ ☐

FISH & ANIMALS:

Feeding
☐

Behavior
☐

Additions
☐

Lost Animals
☐

NOTES

Tank: ----------------------------- **Date:** ---------------

GENERAL CHECK LIST:

Check Equipment ☐   Remove Debris ☐

Water Test ☐   Trim Plants ☐

Rinse Filter ☐   Water Change | 0% | 15% | 25% | 35% | 50% |

WATER PARAMETERS:

Temperature__ ☐

pH_____ ☐

KG_____ ☐

GH_____ ☐

Ammonia____ ☐

Nitrite_____ ☐

Nitrate_____ ☐

Phosphate____ ☐

FISH & ANIMALS:

Feeding

Behavior

Additions

Lost Animals

NOTES

Tank: **Date:**

GENERAL CHECK LIST:

Check Equipment ☐   Remove Debris ☐

Water Test ☐   Trim Plants ☐

Rinse Filter ☐   Water Change | 0% | 15% | 25% | 35% | 50% |

WATER PARAMETERS:

Temperature__ ☐

pH_____ ☐

KG_____ ☐

GH_____ ☐

Ammonia____ ☐

Nitrite_____ ☐

Nitrate_____ ☐

Phosphate____ ☐

FISH & ANIMALS:

Feeding

Behavior

Additions

Lost Animals

NOTES

Tank: _____ **Date:** _____

GENERAL CHECK LIST:

Check Equipment ☐   Remove Debris ☐

Water Test ☐   Trim Plants ☐

Rinse Filter ☐   Water Change | 0% | 15% | 25% | 35% | 50% |

WATER PARAMETERS:

Temperature __ ☐

pH _____ ☐

KG _____ ☐

GH _____ ☐

Ammonia ____ ☐

Nitrite _____ ☐

Nitrate _____ ☐

Phosphate ____ ☐

FISH & ANIMALS:

Feeding

Behavior

Additions

Lost Animals

NOTES

Tank: _____ **Date:** _____

GENERAL CHECK LIST:

Check Equipment [ ]   Remove Debris [ ]

Water Test [ ]   Trim Plants [ ]

Rinse Filter [ ]   Water Change [0%] [15%] [25%] [35%] [50%]

WATER PARAMETERS:

Temperature__ [   ]

pH_____ [   ]

KG_____ [   ]

GH_____ [   ]

Ammonia____ [   ]

Nitrite_____ [   ]

Nitrate_____ [   ]

Phosphate____ [   ]

FISH & ANIMALS:

Feeding
[   ]

Behavior
[   ]

Additions
[   ]

Lost Animals
[   ]

NOTES

Tank: **Date:**

GENERAL CHECK LIST:

Check Equipment ☐   Remove Debris ☐

Water Test ☐   Trim Plants ☐

Rinse Filter ☐   Water Change | 0% | 15% | 25% | 35% | 50% |

WATER PARAMETERS:

Temperature__ ☐

pH_____ ☐

KG_____ ☐

GH_____ ☐

Ammonia____ ☐

Nitrite_____ ☐

Nitrate_____ ☐

Phosphate____ ☐

FISH & ANIMALS:

Feeding

☐

Behavior

☐

Additions

☐

Lost Animals

☐

NOTES

☐

Tank: **Date:**

GENERAL CHECK LIST:

Check Equipment ☐   Remove Debris ☐

Water Test ☐   Trim Plants ☐

Rinse Filter ☐   Water Change | 0% | 15% | 25% | 35% | 50% |

WATER PARAMETERS:

Temperature__ ☐

pH_____ ☐

KG_____ ☐

GH_____ ☐

Ammonia____ ☐

Nitrite_____ ☐

Nitrate_____ ☐

Phosphate____ ☐

FISH & ANIMALS:

Feeding
☐

Behavior
☐

Additions
☐

Lost Animals
☐

NOTES

Tank: **Date:**

GENERAL CHECK LIST:

Check Equipment ☐   Remove Debris ☐

Water Test ☐   Trim Plants ☐

Rinse Filter ☐   Water Change | 0% | 15% | 25% | 35% | 50% |

WATER PARAMETERS:

Temperature _ _ ☐

pH _ _ _ _ _ _ _ _ _ ☐

KG _ _ _ _ _ _ _ _ _ ☐

GH _ _ _ _ _ _ _ _ _ ☐

Ammonia _ _ _ _ ☐

Nitrite _ _ _ _ _ _ ☐

Nitrate _ _ _ _ _ _ ☐

Phosphate _ _ _ _ ☐

FISH & ANIMALS:

Feeding

Behavior

Additions

Lost Animals

NOTES

Tank: _____ **Date:** _____

GENERAL CHECK LIST:

Check Equipment ☐   Remove Debris ☐

Water Test ☐   Trim Plants ☐

Rinse Filter ☐   Water Change ☐ 0% | 15% | 25% | 35% | 50%

WATER PARAMETERS:

Temperature__ ☐

pH_____ ☐

KG_____ ☐

GH_____ ☐

Ammonia____ ☐

Nitrite_____ ☐

Nitrate_____ ☐

Phosphate____ ☐

FISH & ANIMALS:

Feeding

Behavior

Additions

Lost Animals

NOTES

Tank: _____   **Date:** _____

GENERAL CHECK LIST:

Check Equipment ☐   Remove Debris ☐

Water Test ☐   Trim Plants ☐

Rinse Filter ☐   Water Change | 0% | 15% | 25% | 35% | 50% |

WATER PARAMETERS:

Temperature__ ☐

pH_____ ☐

KG_____ ☐

GH_____ ☐

Ammonia____ ☐

Nitrite_____ ☐

Nitrate_____ ☐

Phosphate____ ☐

FISH & ANIMALS:

Feeding
☐

Behavior
☐

Additions
☐

Lost Animals
☐

NOTES

Tank: _____   **Date:** _____

GENERAL CHECK LIST:

Check Equipment ☐   Remove Debris ☐

Water Test ☐   Trim Plants ☐

Rinse Filter ☐   Water Change | 0% | 15% | 25% | 35% | 50% |

WATER PARAMETERS:

Temperature__ ☐

pH_____ ☐

KG_____ ☐

GH_____ ☐

Ammonia____ ☐

Nitrite_____ ☐

Nitrate_____ ☐

Phosphate____ ☐

FISH & ANIMALS:

Feeding

Behavior

Additions

Lost Animals

NOTES

Tank: **Date:**

GENERAL CHECK LIST:

Check Equipment ☐   Remove Debris ☐

Water Test ☐   Trim Plants ☐

Rinse Filter ☐   Water Change | 0% | 15% | 25% | 35% | 50% |

WATER PARAMETERS:

Temperature __ ☐

pH _____ ☐

KG _____ ☐

GH _____ ☐

Ammonia ____ ☐

Nitrite _____ ☐

Nitrate _____ ☐

Phosphate ____ ☐

FISH & ANIMALS:

Feeding

Behavior

Additions

Lost Animals

NOTES

Tank: _____   **Date:** _____

GENERAL CHECK LIST:

Check Equipment ☐   Remove Debris ☐

Water Test ☐   Trim Plants ☐

Rinse Filter ☐   Water Change | 0% | 15% | 25% | 35% | 50% |

WATER PARAMETERS:

Temperature __ ☐

pH _____ ☐

KG _____ ☐

GH _____ ☐

Ammonia ____ ☐

Nitrite _____ ☐

Nitrate _____ ☐

Phosphate ____ ☐

FISH & ANIMALS:

Feeding
☐

Behavior
☐

Additions
☐

Lost Animals
☐

NOTES

Tank: **Date:**

GENERAL CHECK LIST:

Check Equipment ☐   Remove Debris ☐

Water Test ☐   Trim Plants ☐

Rinse Filter ☐   Water Change | 0% | 15% | 25% | 35% | 50% |

WATER PARAMETERS:

Temperature __ ☐

pH _____ ☐

KG _____ ☐

GH _____ ☐

Ammonia ____ ☐

Nitrite _____ ☐

Nitrate _____ ☐

Phosphate ____ ☐

FISH & ANIMALS:

Feeding

Behavior

Additions

Lost Animals

NOTES

Tank: _____   **Date:** _____

GENERAL CHECK LIST:

Check Equipment ☐   Remove Debris ☐

Water Test ☐   Trim Plants ☐

Rinse Filter ☐   Water Change | 0% | 15% | 25% | 35% | 50% |

WATER PARAMETERS:

Temperature __ ☐

pH _____ ☐

KG _____ ☐

GH _____ ☐

Ammonia ____ ☐

Nitrite _____ ☐

Nitrate _____ ☐

Phosphate ____ ☐

FISH & ANIMALS:

Feeding
☐

Behavior
☐

Additions
☐

Lost Animals
☐

NOTES

Tank: _____   **Date:** _____

GENERAL CHECK LIST:

Check Equipment ☐   Remove Debris ☐

Water Test ☐   Trim Plants ☐

Rinse Filter ☐   Water Change | 0% | 15% | 25% | 35% | 50% |

WATER PARAMETERS:

Temperature_ _ ☐

pH_ _ _ _ _ _ _ _ ☐

KG_ _ _ _ _ _ _ _ ☐

GH_ _ _ _ _ _ _ _ ☐

Ammonia_ _ _ _ ☐

Nitrite_ _ _ _ _ _ ☐

Nitrate_ _ _ _ _ _ ☐

Phosphate_ _ _ _ ☐

FISH & ANIMALS:

Feeding
☐

Behavior
☐

Additions
☐

Lost Animals
☐

NOTES

Tank: **Date:**

GENERAL CHECK LIST:

Check Equipment ☐   Remove Debris ☐

Water Test ☐   Trim Plants ☐

Rinse Filter ☐   Water Change | 0% | 15% | 25% | 35% | 50% |

WATER PARAMETERS:

Temperature__ ☐

pH_____ ☐

KG_____ ☐

GH_____ ☐

Ammonia____ ☐

Nitrite_____ ☐

Nitrate_____ ☐

Phosphate____ ☐

FISH & ANIMALS:

Feeding

Behavior

Additions

Lost Animals

NOTES

Tank: **Date:**

GENERAL CHECK LIST:

Check Equipment ☐   Remove Debris ☐

Water Test ☐   Trim Plants ☐

Rinse Filter ☐   Water Change | 0% | 15% | 25% | 35% | 50% |

WATER PARAMETERS:

Temperature__ ☐

pH_____ ☐

KG_____ ☐

GH_____ ☐

Ammonia____ ☐

Nitrite_____ ☐

Nitrate_____ ☐

Phosphate____ ☐

FISH & ANIMALS:

Feeding

Behavior

Additions

Lost Animals

NOTES

Tank: **Date:**

GENERAL CHECK LIST:

Check Equipment ☐   Remove Debris ☐

Water Test ☐   Trim Plants ☐

Rinse Filter ☐   Water Change | 0% | 15% | 25% | 35% | 50% |

WATER PARAMETERS:

Temperature __ ☐

pH _____ ☐

KG _____ ☐

GH _____ ☐

Ammonia ____ ☐

Nitrite _____ ☐

Nitrate _____ ☐

Phosphate ____ ☐

FISH & ANIMALS:

Feeding
☐

Behavior
☐

Additions
☐

Lost Animals
☐

NOTES

Tank: **Date:**

GENERAL CHECK LIST:

Check Equipment ☐   Remove Debris ☐

Water Test ☐   Trim Plants ☐

Rinse Filter ☐   Water Change | 0% | 15% | 25% | 35% | 50% |

WATER PARAMETERS:

Temperature __ ☐

pH _____ ☐

KG _____ ☐

GH _____ ☐

Ammonia ____ ☐

Nitrite _____ ☐

Nitrate _____ ☐

Phosphate ____ ☐

FISH & ANIMALS:

Feeding

Behavior

Additions

Lost Animals

NOTES

Tank: **Date:**

GENERAL CHECK LIST:

Check Equipment ☐   Remove Debris ☐

Water Test ☐   Trim Plants ☐

Rinse Filter ☐   Water Change | 0% | 15% | 25% | 35% | 50% |

WATER PARAMETERS:

Temperature__ ☐

pH_____ ☐

KG_____ ☐

GH_____ ☐

Ammonia____ ☐

Nitrite_____ ☐

Nitrate_____ ☐

Phosphate____ ☐

FISH & ANIMALS:

Feeding

Behavior

Additions

Lost Animals

NOTES

Tank: **Date:**

GENERAL CHECK LIST:

Check Equipment ☐   Remove Debris ☐

Water Test ☐   Trim Plants ☐

Rinse Filter ☐   Water Change | 0% | 15% | 25% | 35% | 50% |

WATER PARAMETERS:

Temperature__ ☐

pH_____ ☐

KG_____ ☐

GH_____ ☐

Ammonia____ ☐

Nitrite_____ ☐

Nitrate_____ ☐

Phosphate____ ☐

FISH & ANIMALS:

Feeding

Behavior

Additions

Lost Animals

NOTES

Tank: _____   **Date:** _____

GENERAL CHECK LIST:

Check Equipment ☐   Remove Debris ☐

Water Test ☐   Trim Plants ☐

Rinse Filter ☐   Water Change | 0% | 15% | 25% | 35% | 50% |

WATER PARAMETERS:

Temperature__ ☐

pH_____ ☐

KG_____ ☐

GH_____ ☐

Ammonia____ ☐

Nitrite_____ ☐

Nitrate_____ ☐

Phosphate____ ☐

FISH & ANIMALS:

Feeding

Behavior

Additions

Lost Animals

NOTES

Tank: **Date:**

GENERAL CHECK LIST:

Check Equipment ☐   Remove Debris ☐

Water Test ☐   Trim Plants ☐

Rinse Filter ☐   Water Change | 0% | 15% | 25% | 35% | 50% |

WATER PARAMETERS:

Temperature __ ☐

pH _____ ☐

KG _____ ☐

GH _____ ☐

Ammonia ____ ☐

Nitrite _____ ☐

Nitrate _____ ☐

Phosphate ____ ☐

FISH & ANIMALS:

Feeding

Behavior

Additions

Lost Animals

NOTES

Tank: _____ **Date:** _____

Check Equipment ☐   Remove Debris ☐

Water Test ☐   Trim Plants ☐

Rinse Filter ☐   Water Change | 0% | 15% | 25% | 35% | 50% |

WATER PARAMETERS:

Temperature __ ☐

pH _____ ☐

KG _____ ☐

GH _____ ☐

Ammonia ____ ☐

Nitrite _____ ☐

Nitrate _____ ☐

Phosphate ____ ☐

FISH & ANIMALS:

Feeding

☐

Behavior

☐

Additions

☐

Lost Animals

☐

NOTES

☐

Tank: **Date:**

GENERAL CHECK LIST:

Check Equipment ☐   Remove Debris ☐

Water Test ☐   Trim Plants ☐

Rinse Filter ☐   Water Change 0% | 15% | 25% | 35% | 50%

WATER PARAMETERS:

Temperature__ ☐

pH_____ ☐

KG_____ ☐

GH_____ ☐

Ammonia____ ☐

Nitrite_____ ☐

Nitrate_____ ☐

Phosphate____ ☐

FISH & ANIMALS:

Feeding

Behavior

Additions

Lost Animals

NOTES

Tank: _____   **Date:** _____

GENERAL CHECK LIST:

Check Equipment ☐   Remove Debris ☐

Water Test ☐   Trim Plants ☐

Rinse Filter ☐   Water Change | 0% | 15% | 25% | 35% | 50% |

WATER PARAMETERS:

Temperature __ ☐

pH _____ ☐

KG _____ ☐

GH _____ ☐

Ammonia ____ ☐

Nitrite _____ ☐

Nitrate _____ ☐

Phosphate ____ ☐

FISH & ANIMALS:

Feeding

Behavior

Additions

Lost Animals

NOTES

Tank: **Date:**

GENERAL CHECK LIST:

Check Equipment ☐   Remove Debris ☐

Water Test ☐   Trim Plants ☐

Rinse Filter ☐   Water Change | 0% | 15% | 25% | 35% | 50% |

WATER PARAMETERS:

Temperature __ ☐

pH _____ ☐

KG _____ ☐

GH _____ ☐

Ammonia ____ ☐

Nitrite _____ ☐

Nitrate _____ ☐

Phosphate ____ ☐

FISH & ANIMALS:

Feeding

Behavior

Additions

Lost Animals

NOTES

Tank: _____ **Date:** _____

Check Equipment ☐   Remove Debris ☐

Water Test ☐   Trim Plants ☐

Rinse Filter ☐   Water Change | 0% | 15% | 25% | 35% | 50% |

WATER PARAMETERS:

Temperature__ ☐

pH_____ ☐

KG_____ ☐

GH_____ ☐

Ammonia____ ☐

Nitrite_____ ☐

Nitrate_____ ☐

Phosphate____ ☐

FISH & ANIMALS:

Feeding

Behavior

Additions

Lost Animals

NOTES

Tank: **Date:**

GENERAL CHECK LIST:

Check Equipment ☐   Remove Debris ☐

Water Test ☐   Trim Plants ☐

Rinse Filter ☐   Water Change | 0% | 15% | 25% | 35% | 50% |

WATER PARAMETERS:

Temperature __ ☐

pH _____ ☐

KG _____ ☐

GH _____ ☐

Ammonia ____ ☐

Nitrite _____ ☐

Nitrate _____ ☐

Phosphate ____ ☐

FISH & ANIMALS:

Feeding

Behavior

Additions

Lost Animals

NOTES

Tank: _____   **Date:** _____

GENERAL CHECK LIST:

Check Equipment ☐   Remove Debris ☐

Water Test ☐   Trim Plants ☐

Rinse Filter ☐   Water Change | 0% | 15% | 25% | 35% | 50% |

WATER PARAMETERS:

Temperature __ ☐

pH _____ ☐

KG _____ ☐

GH _____ ☐

Ammonia ____ ☐

Nitrite _____ ☐

Nitrate _____ ☐

Phosphate ____ ☐

FISH & ANIMALS:

Feeding
☐

Behavior
☐

Additions
☐

Lost Animals
☐

NOTES

Tank: **Date:**

GENERAL CHECK LIST:

Check Equipment ☐   Remove Debris ☐

Water Test ☐   Trim Plants ☐

Rinse Filter ☐   Water Change | 0% | 15% | 25% | 35% | 50% |

WATER PARAMETERS:

Temperature _ _ ☐

pH _ _ _ _ _ _ _ _ ☐

KG _ _ _ _ _ _ _ _ ☐

GH _ _ _ _ _ _ _ _ ☐

Ammonia _ _ _ _ ☐

Nitrite _ _ _ _ _ _ ☐

Nitrate _ _ _ _ _ _ ☐

Phosphate _ _ _ _ ☐

FISH & ANIMALS:

Feeding

Behavior

Additions

Lost Animals

NOTES

Tank: _____ **Date:** _____

GENERAL CHECK LIST:

Check Equipment ☐ Remove Debris ☐

Water Test ☐ Trim Plants ☐

Rinse Filter ☐ Water Change [0%] [15%] [25%] [35%] [50%]

WATER PARAMETERS:

Temperature__ ☐

pH_____ ☐

KG_____ ☐

GH_____ ☐

Ammonia____ ☐

Nitrite_____ ☐

Nitrate_____ ☐

Phosphate____ ☐

FISH & ANIMALS:

Feeding
☐

Behavior
☐

Additions
☐

Lost Animals
☐

NOTES

Thank You

Owning an aquarium is an incredibly enriching experience. They mesmerize viewers with their gorgeous inhabitants and captivate with their tranquility. Thank you for *diving* into this fantastic hobby and learning more about taking care of these funny guys and girls. Enjoy taking care of your fish roommates, and I wish you much success and joy.

If you've enjoyed this book, **please let me know by leaving an Amazon rating and a brief review!** It only takes about 30 seconds, and it helps me compete against big publishing houses. It also helps other readers find my work! Thank you for your time, and have an awesome time caring for your flappy friend(s).

Cheers,

Walter

Made in the USA
Columbia, SC
21 February 2022

56580354R00065